EMC Pocket Guide

Key EMC Facts, Equations and Data

Kenneth Wyatt & Randy Jost, Ph.D.

T0188377

SciTech
PUBLISHING
an imprint of the IET

Published by SciTech Publishing, an imprint of the IET.
www.scitechpub.com
www.theiet.org

This book is available at special quantity discounts to use as
premiums and sales promotions, or for use in corporate
training programs. For more information and quotes, email
marketing@scitechpub.com.

10 9 8 7 6 5 4 3 2 1

ISBN 978-1-61353-191-4
ISBN 978-1-61353-192-1 (PDF)

Typeset in India by MPS Ltd
Printed in the USA by Docusource (Raleigh, NC)

Contents

■ EMC Fundamentals

What is EMC?

Electromagnetic Compatibility is achieved when:

- Electronic products do not interfere with their environments (emissions)
- The environments do not upset the operation of electronic products (immunity)
- The electronic product does not interfere with itself (signal integrity)

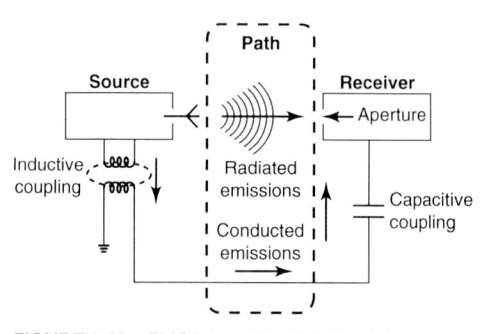

FIGURE 1 Key EMC Interaction Relationships.

In reviewing the various ways signals can be propagated within and between systems, we see that energy is transferred from source to receiver (victim) via some coupling path. The energy can be propagated on transmission or power line connections between various systems or subsystems (***conducted emissions***), with the coupling of energy between source and conductor facilitated by inductive or capacitive means of coupling. Another pathway is by radiation of electromagnetic waves from source to receiver (***radiated emissions***) or a combination of both ways.

The most common EMC issue is radiated emissions (RE). In order to have RE, the system must be comprised of two antennas – one for the source of energy and one for the receptor, or victim. The latter is generally the EMI receiver or spectrum analyzer.

In order for the system under test to emit RF radiation (RE), there must be a source of energy and an antenna. If there's no energy source, there can be no emissions and, likewise, if there is no antenna, there can be no emissions.

Thus, you can "source suppress" the energy (best policy) by slowing down faster than required edges (through low-pass filtering), re-routing clock lines and other high speed traces to shorten them and avoid running them across gaps in return planes or changing reference planes without clearly defined return paths.

You can eliminate the antenna by blocking the coupling path from the energy source (shielding or filtering, for example) or through techniques described in "Hidden Antennas" (page 32 below).

Frequency versus Time Domain

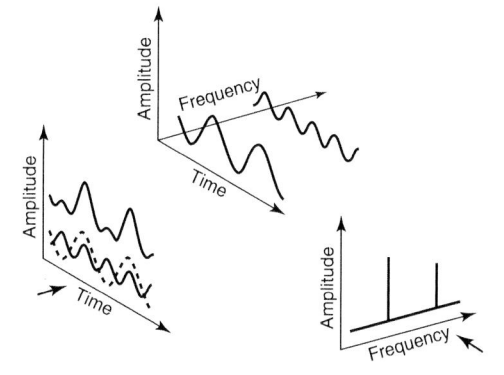

FIGURE 2 The Relationship of Time Domain to Frequency Domain.

Both oscilloscopes (time domain) and spectrum analyzers (frequency domain) are useful for identifying, characterizing or troubleshooting EMC issues in a product.

Very often, the issues are frequency-related and are best identified using a spectrum analyzer.

According to the emission standards, be sure to set the resolution bandwidth to 9 kHz (or 10 kHz*) for frequencies below 30 MHz and 120 kHz (or 100 kHz*) for frequencies from 30 to 1000 MHz and 1 MHz for frequencies 1 GHz and higher. Set the video bandwidth the same, or wider for best amplitude accuracy.

Oscilloscopes are very useful for identifying resonances or ringing on digital or clock signals, as well as crosstalk issues.

*for analyzers lacking the "EMI bandwidths".

Fourier Series & Transforms

Fourier analysis is a key tool in understanding the signals that are encountered in EMC problems and solution. Fourier series help provide a discrete representation of time domain signals, while Fourier transforms are normally used to analyze continuous signals.

Signal Spectra

The envelopes of the magnitude spectrum of rectangular and trapezoidal pulses provide insight into the spectral content of the waveforms typically encountered in EMC work. Additionally, pulse widths and duty cycles have a strong influence on spectral content and amplitudes. Knowing these factors, we can estimate what harmonics we need to protect against and at what levels.

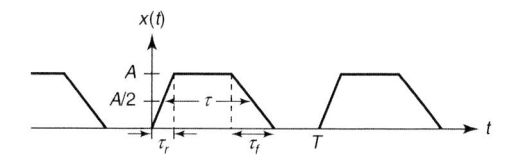

FIGURE 3 A typical trapezoidal waveform.

Figure 3 shows a trapezoidal waveform that represents the output of the clock for digital circuits. By selecting the duty cycle, τ, the rise time, τ_r, and fall time, τ_f, appropriately, we can tailor our waveform to meet required performance criteria, while minimizing the negative impact of higher order harmonics on adjacent analog and digital circuits.

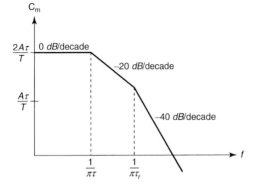

FIGURE 4 Spectral bounds on trapezoidal waveform.

Referring to Figure 4, note that the duty cycle, τ, determines the first breakpoint, while the rise time, τ_r, determines the second breakpoint. Figure 5 illustrates that the slower the rise time, the lower the magnitude of higher order harmonics.

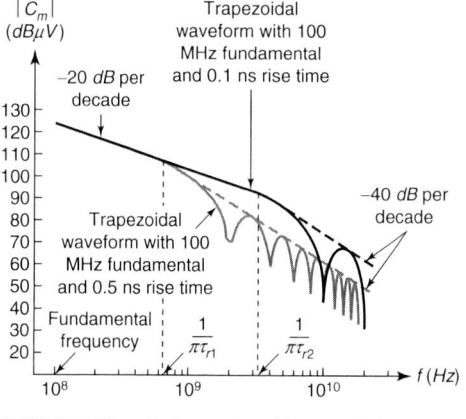

FIGURE 5 Magnitude spectra of trapezoidal waveforms vs. rise times.

Figure 6 illustrates that reducing the duty cycle will decrease the level of the magnitude envelope, again decreasing the level of higher order harmonics.

In applying these principles, each individual situation will have to be evaluated. However, in general, there will be a −20 dB per decade drop in the spectral magnitude after the first breakpoint, which will occur at $^1/_{\pi\tau}$, and then a drop of −40 dB per decade after the second breakpoint,

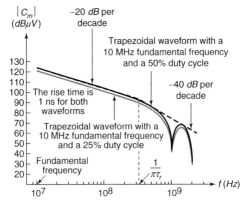

FIGURE 6 Magnitude spectra of trapezoidal waveforms vs. duty cycle.

generally occurring at $1/\pi\tau_r$. Thus paying attention to both the rise/fall time and duty cycle will pay significant dividends in controlling the harmonic content of clocks and oscillator circuits.

■ EMC Design

Cable Terminations

Cable shields should be connected (bonded) very well to the metallic enclosure. If the shield connection is made directly to the PC board,

---

common-mode noise currents (I_{cm}) can flow directly through the enclosure and out the I/O or power cable.

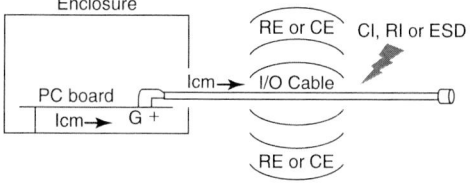

FIGURE 7

If the cable shield is connected in one place on the shielded enclosure, that's called a "pigtail" connection and is nearly as bad. Ideally, connect the cable shield in multiple places (360°).

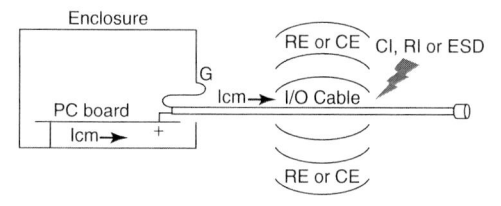

FIGURE 8

A proper 360-degree bond to enclosure solves many EMC issues.

Connector Bonding

Connector shields must be bonded in multiple places to the shielded enclosure. If this is not done, then high-frequency common-mode currents generated on the PC board may flow out the I/O or power cables and cause radiated emissions. Equally, cables may pick up outside radiated or conducted sources, which can cause circuit upsets.

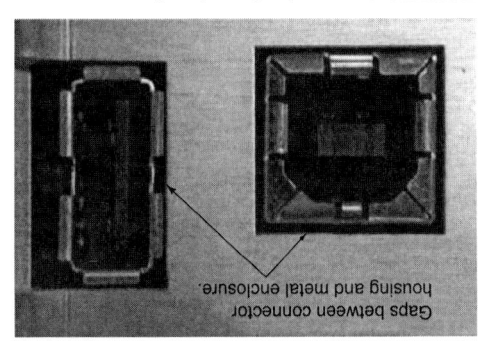

Gaps between connector housing and metal enclosure.

FIGURE 9 An example of poor bonding.

If connectors are unshielded, then all signal, return and power signals must be filtered on the PC board right at the connector. In many cases, a metal plate mounted closely under the PC board with one end turned up 90-degrees and bonded well to all the I/O connector ground shells can provide some protection. Note that the bonding must be designed to deal with the harshest environment to which the system will be exposed.

Shielding

Never penetrate a shield with a wire or cable without using some form of cable shielding or filtering. Common-mode noise currents on the PC board or internal electronic sub-assemblies will couple to the wire or cable and travel outside, causing radiated or conducted emissions, or, conversely, susceptibility to external radiated or conducted fields.

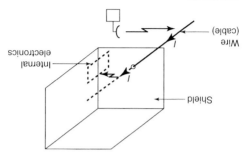

FIGURE 10

Calculating Shielding Effectiveness (SE):

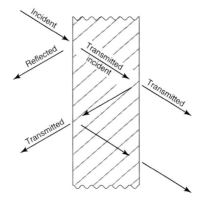

FIGURE 11

$$SE(dB) = A(dB) + R(dB) + M(dB)$$

Absorption Loss

$$A_{dB} = 20 \log_{10} e^{t/\delta}$$

$$A_{dB} = 20t/\delta \log_{10} e$$

$$A_{dB} = 8.6859t/\delta$$

$$A_{dB} = 131.4t \sqrt{f \mu_r \sigma_r} \quad \text{(where } t \text{ is in m)}$$

$$A_{dB} = 3.338t \sqrt{f \mu_r \sigma_r} \quad \text{(where } t \text{ is in inches)}$$

Note that for every skin depth, δ, $A = 8.6859$ dB

Reflection Loss

$$R_{dB} = 168 + 10 \log_{10} \left(\frac{\sigma_r}{\mu_r f} \right)$$

Multiple Reflection Loss

$$M_{dB} = 20 \log_{10} \left| 1 - \left(\frac{\eta_0 - \hat{\eta}}{\eta_0 + \hat{\eta}} \right)^2 e^{-2t/\delta} e^{-j2\beta t} \right|$$

$$M_{dB} \cong 20 \log_{10} \left| 1 - e^{2t/\delta} e^{-j2t/\delta} \right|$$

For most good shields, M is zero.

SE versus Slot Length

Freq (MHz)	20 dB SE	40 dB SE
10	100 cm	19 cm
30	75 cm	5 cm
100	15 cm	1.5 cm
300	5 cm	0.5 cm
500	2.5 cm	—
1000	1.5 cm	—

For example, a 6-inch slot only has an effective shielding effectiveness (SE) of 20 dB at 100 MHz. Harmonic frequencies not on the chart may be interpolated.

Most practical metallic enclosures (with required apertures, realistic seams, and ventilation holes) averages 20 to 30 dB SE. As the frequency increases, it becomes increasingly important to ensure any holes and seams are minimized in your enclosure.

Leakages may be measured with either E-field or H-field probes. The length of major leakages should be marked at the end points and then analyzed for potential resonances at critical harmonic frequencies using the formulas on page 32 or the Frequency versus Wavelength table on page 34. For example, a seam with leakage 15 cm long is 1/4-wavelength at 500 MHz. For a 500 MHz harmonic, you'd want no more than 2.5 cm leakage for 20 dB SE.

Common PC Board Issues

Discontinuous Return Paths

Most PC board problems can be traced to discontinuous signal return paths. This is becoming more of an issue with the increasing clock frequencies used today. This is also a major cause of common-mode emissions.

1. High-frequency signals travel down the trace to the load and then return immediately under that trace, due to mutual coupling.

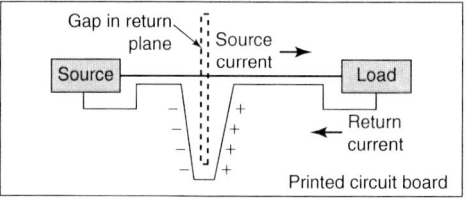

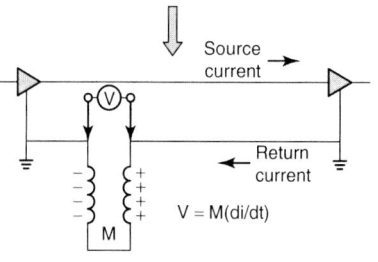

FIGURE 12

    a. All too often that return path is interrupted by a discontinuity, such as a gap or slot in the return or power planes (Figure 12).

    b. *Solution*: Examine the signal return and power plane layers for gaps and slots.

2. The signal trace passes through a via and changes reference planes.

    a. Add extra vias or stitching capacitors for return currents when switching reference planes (Figure 13).

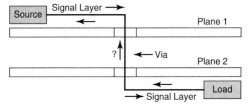

FIGURE 13

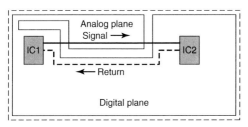

FIGURE 14

3. Routing a clock (or any) trace over an unrelated (e.g., analog) plane can cause noise coupling to sensitive circuitry (Figure 14).
   a. High-frequency clock signals can couple into the analog circuitry (crosstalk).
   b. Because the return path is now forced out around the unrelated plane, this causes common-mode currents and resulting radiated emissions.

Effects of ESD / Preventing ESD Problems

Two primary effects of an ESD event:

1. An intense electrostatic field created by the charge separation prior to the ESD arc. It can overstress dielectric materials.
2. An intense arc discharge.

The arc discharge can cause the following:

1. Direct conduction through the electronic circuitry (destroys devices).
2. Secondary arcs or discharges (near field E/H fields).
3. Capacitive coupling to circuits (high-impedance circuits).
4. Inductive coupling to circuits (low-impedance circuits).

Four ways to prevent problems caused by an ESD event:

1. Prevent the occurrence of the ESD event (through control of the environment).
2. Prevent or reduce the coupling.
3. Add a low-pass filter (R-C) to processor reset pins.
4. Create an inherent immunity through software.
   a. Don't use unlimited "wait states".

b. Use "watchdog" routines to restart a product.

c. Use parity bits, checksums, or error correcting codes to prevent storage of bad data.

d. All inputs should be latched and strobed – don't leave any floating.

■ EMC Measurements

Troubleshooting with Current Probes

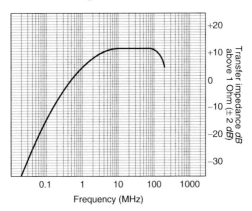

FIGURE 15 A plot of transfer impedance for a typical current probe (Fischer Custom Communications F-33-1 probe used).

By clamping a current probe around a cable and using the plot of transfer impedance, you can calculate the common-mode current in that cable versus frequency. By knowing the current, you can plug it into the equation for the E-field $(dB\mu V/m)$ and compare to the radiated emission limit. See "Commonly Used Equations" (page 38).

$$I_{cm}(dB\mu A) = V_{term}(dB\mu V) - Zt(dB\Omega)$$

Troubleshooting with Antennas

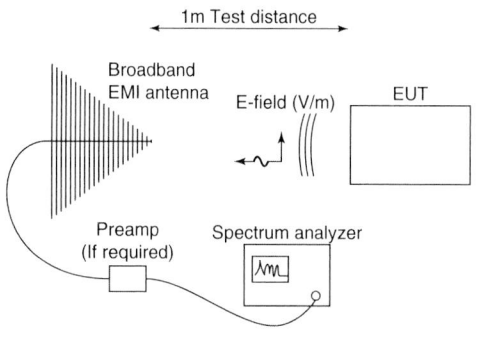

FIGURE 16 Setup for radiated emissions troubleshooting.

To troubleshoot radiated emissions, you'll need to find some space at least 3 m from the nearest reflective structure, such as a parking lot, large conference room, or basement.

Set up the antenna about 1 m from the EUT and connect the antenna to a spectrum analyzer, via a broadband preamp if the signal requires boosting. You'll need to take the 3 m (or 10 m) limits and add 10 dB (or 20 dB) to them to convert to 1 m test distance limits (see Figure 17). Because you're partly measuring in the near field, these computed test limits may not be completely accurate but should at least indicate major "red flags". The antenna manufacturer will provide a chart of Antenna Factors (AF).

If using a log-periodic antenna, understand the active element will change position with frequency: higher frequencies will be closer than 1 m, lower frequencies farther.

Use these extrapolated FCC/CISPR radiated emissions limits when troubleshooting with an antenna positioned 1 m away from the EUT.

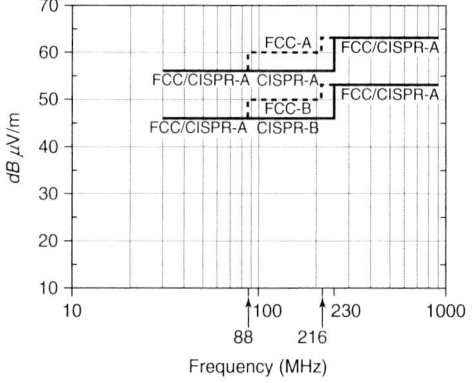

FIGURE 17 Radiated emission limits extrapolated to a 1 m test distance.

Use this formula to determine the E-field $(dB\mu V/m)$ at the antenna terminals:

$$E\text{-Field }(dB\mu V/m) = SpecAnalyzer(dB\mu V)$$
$$- PreampGain(dB) + CoaxLoss(dB)$$
$$+ AttenuatorLoss(dB) + AntFactor(dB)$$

■ EMC Standards

Common Basic EMC Standards

Type	Standard	Test/Scope
Commercial	CISPR 11	ISM Equipment
	CISPR 22	ITE Equipment
	CISPR 16	Methods of Measurement
	IEC 61000-3-2	Harmonics
	IEC 61000-3-3	Flicker
	IEC 61000-4-2	Electrostatic Discharge (ESD)
	IEC 61000-4-3	Radiated Immunity
	IEC 61000-4-4	Electrically Fast Transient (EFT)
	IEC 61000-4-5	Surge (Lightning)
	IEC 61000-4-6	Conducted Immunity
	IEC 61000-4-8	Magnetic Immunity
	IEC 61000-4-11	Dips, Interrupts, Voltage Variations
	FCC Part 15B	ITE Equipment
	ANSI C63.4	Methods of Measurement
Medical	IEC 60601-1-2	Medical Products
Automotive	SAE J1113	Automotive EMC
Military	MIL STD 461F	EMC Test Requirements
Aerospace	DO-160	EMC Test Requirements (Aircraft)
	SAE ARP5412B	Aircraft Lightning Environment & Related Test Waveforms
	SAE ARP5416A	Aircraft Lightning Test Methods

In general, the emission limits are more restrictive for residential environments and less restrictive for industrial environments. Immunity, on the other hand, is usually more restrictive for industrial environments.

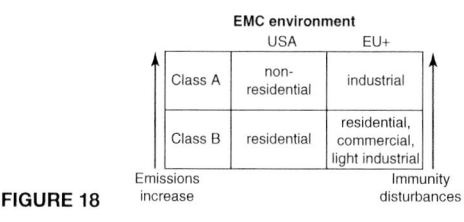

EMC environment

	USA	EU+
Class A	non-residential	industrial
Class B	residential	residential, commercial, light industrial

Emissions increase ↑   ↑ Immunity disturbances

FIGURE 18

FCC/CISPR 10 m Limits (RE)

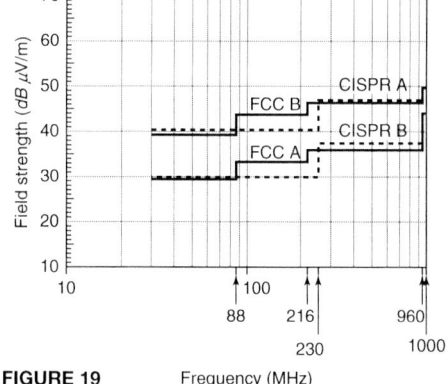

FIGURE 19   Frequency (MHz)

FCC/CISPR Limits (CE)

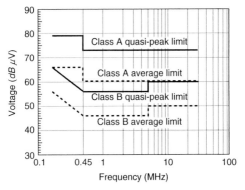

FIGURE 20

Class A – Industrial/commercial environment

Class B – Residential environment

■ Using Decibels

Working with dB

The decibel is always a ratio...

- Power Gain = P_{out}/P_{in}
- Power Gain (dB) = $10 \log (P_{out}/P_{in})$
- Voltage Gain (dB) = $20 \log (V_{out}/V_{in})$
- Current Gain (dB) = $20 \log (I_{out}/I_{in})$

We commonly work with...

- dBm (referenced to 1 mW)
- dBμV (referenced to 1 μV)
- dBμA (referenced to 1 μA)

Power Ratios

3 dB = double (or half) the power

10 dB = 10X (or /10) the power

Voltage/Current Ratios

6 dB = double (or half) the voltage/current

20 dB − 10X (or /10) the voltage/current

dBm, dBμV, dBμA (conversion)

Log ↔ Linear Voltage

Volts to dBV	$dBV = 20\log(V)$
Volts to dBμV	$dB\mu V = 20\log(V) + 120$
dBV to Volts	$V = 10^{(dBV/20)}$
dBμV to Volts	$V = 10^{((dB\mu V-120)/20)}$
dBV to dBμV	$dB\mu V = dBV + 120$
dBμV to dBV	$dBV = dB\mu V - 120$

Note: For current relationships, substitute A for V

Log identities:

If $Y = \log(X)$, then $X = 10^Y$

$\log 1 = 0$

Log of numbers > 1 are positive

Log of numbers < 1 are negative

The log $(A*B) = \log A + \log B$

The log $(A/B) = \log A - \log B$

The log of $A^n = n*\log A$

dBm to dBμV

(Assuming a 50-Ohm system)

dBm	dBμV
20	127
10	117
0	107
−10	97
−20	87
−30	77
−40	67
−50	57
−60	47
−70	37
−80	27
−90	17
−100	7

To convert dBm to dBμV, use:

$$dB\mu V = dBm + 107$$

Power Ratios (dB)

Unit	Power	Voltage or Current
0.1	−10 dB	−20 dB
0.2	−7.0 dB	−14.0 dB
0.3	−5.2 dB	−10.5 dB
0.5	−3.0 dB	−6.0 dB
1	0 dB	0 dB
2	3.0 dB	6.0 dB
3	4.8 dB	9.5 dB
5	7.0 dB	14.0 dB
7	8.5 dB	16.9 dB
8	9.0 dB	18.1 dB
9	9.5 dB	19.1 dB
10	10 dB	20 dB
20	13.0 dB	26.0 dB
30	14.8 dB	29.5 dB
50	17.0 dB	34.0 dB
100	20 dB	40 dB
1,000	30 dB	60 dB
1,000,000	60 dB	120 dB

Double the value and it adds 3 dB to power, 6 dB to voltage and current. Halving the value subtracts 3 dB from power, 6 dB from voltage and current.

Multiplying 10X the value and it adds 10 dB to power, 20 dB to voltage and current. Dividing by 10X the value subtracts 10 dB from power, 20 dB from voltage and current.

■ Frequency versus Wavelength

Most EMC issues occur in the range 9 kHz through 6 GHz. Conducted emissions tend to occur below 30 MHz and radiated emissions tend to occur above 30 MHz.

The Electromagnetic Spectrum

Band	Freq Range	Wavelength
HF	3–30 MHz	100 m – 10 m
VHF	30–300 MHz	10 m – 1 m
UHF	300–1000 MHz	1 m – 30 cm
L	1–2 GHz	30 cm – 15 cm
S	2–4 GHz	15 cm – 7.5 cm
C	4–8 GHz	7.5 cm – 3.75 cm
X	8–12 GHz	3.75 cm – 2.5 cm
Ku	12–18 GHz	2.5 cm – 1.67 cm
K	18–27 GHz	1.67 cm – 1.11 cm
Ka	27–40 GHz	1.11 cm – 0.75 cm

For more detail on the users of the EM Spectrum review the chart at:

http://www.ntia.doc.gov/files/ntia/publications/ spectrum_wall_chart_aug2011.pdf. Note that this is the spectrum allocation chart for the USA. Other countries may have similar allocation charts.

Hidden Antennas

An important concept to grasp is the electrical dimension of an electromagnetic radiating structure (we sometimes call these "*antennas*"). This is expressed in terms of *wavelength* (λ).

In a lossless medium (free space),

$$\text{Wavelength } (\lambda) = v/f$$

where v = velocity of propagation and f = frequency (Hz).

In free space $v = v_o \approx 3 \times 10^8$ m/s (approx. speed of light)

Easy to remember formulas for wavelength:

$$\lambda(m) = 300/f \,(\text{MHz})$$

$$\lambda/2 \,(ft) = 468/f \,(\text{MHz})$$

This becomes important when it comes to identifying potential radiating structures – so called "hidden antennas" – of the product or system under test. These structures could include:

1. Cables (I/O or power)
2. Seams/slots in shielded enclosures
3. Apertures in enclosures
4. Poorly bonded sheet metal (of enclosures)
5. PC boards (especially narrow ones)
6. Internal interconnect cables
7. Heat sinks
8. Daughter boards
9. Peripheral equipment connected to the EUT
10. Power plane "patch" over a ground plane

For example, as a cable or slot approaches 1/4 wavelength (or a multiple) at the frequency of concern, it becomes an efficient transmitting or receiving antenna. Conversely, 1/20th wavelength makes a poor antenna. Use the following charts as for help.

Frequency versus Wavelength (air)

Freq	Wavelength	1/4	1/20th
10 Hz	30,000 km	7,500 km	1,500 km
60 Hz	5,000 km	1,250 km	250 km
400 Hz	750 km	187.5 km	37.5 km
1 kHz	300 km	125 km	37.5 km
10 kHz	30 km	7.5 km	1.5 km
100 kHz	3 km	750 m	150 m
1 MHz	300 m	75 m	15 m
10 MHz	30 m	7.5 m	1.5 m
100 MHz	3 m	75 cm	15 cm
300 MHz	100 cm	25 cm	5 cm
500 MHz	60 cm	15 cm	3 cm
1 GHz	30 cm	7.5 cm	1.5 cm
10 GHz	3 cm	.75 cm	.15 cm

A slot or cable of 1/4 wavelength (or multiple) can make a good antenna at the frequency of concern.

A slot or cable of 1/20th wavelength makes a very poor antenna at the frequency of concern.

Frequency versus Wavelength (FR-4)

Freq	Wavelength	1/4	1/20th
10 MHz	18 m	4.5 m	90 cm
25 MHz	7.2 m	1.8 m	36 cm
50 MHz	3.6 m	90 cm	18 cm
100 MHz	1.8 m	45 cm	9 cm
200 MHz	90 cm	22.5 cm	4.5 cm
300 MHz	60 cm	15 cm	3 cm
400 MHz	45 cm	11 cm	2.25 cm
500 MHz	36 cm	9 cm	1.8 cm
600 MHz	30 cm	7.5 cm	1.5 cm
700 MHz	25.7 cm	6.4 cm	1.3 cm
800 MHz	22.5 cm	5.6 cm	1.1 cm
900 MHz	20 cm	5 cm	1 cm
1000 MHz	18 cm	4.5 cm	0.9 cm

This is just an approximate length, due to variations in the velocity constant of FR-4 fiberglass-epoxy.

A velocity factor of 0.6 was used in the calculation, but it can vary from 0.4 to 0.8. Basically, a quarter-wave trace or PC board length will be shorter than in free space by the velocity factor.

Ohms Law (formula wheel)

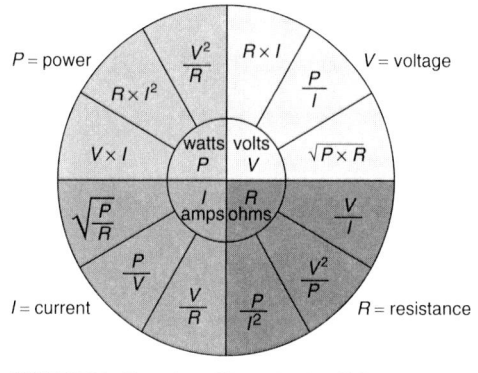

FIGURE 21 Ohms Law "formula wheel" for calculating resistance (R), voltage (V), current (I) or power (P), given at least two of the other values.

VSWR and Return Loss

VSWR given forward/reverse power

$$\text{VSWR} = \frac{1 + \sqrt{\dfrac{P_{\text{rev}}}{P_{\text{fwd}}}}}{1 - \sqrt{\dfrac{P_{\text{rev}}}{P_{\text{fwd}}}}}$$

VSWR given reflection coefficient

$$\text{VSWR} = \frac{1 + \rho}{1 - \rho}$$

Reflection coefficient, ρ, given Z_1, Z_2 Ohms

$$\rho = \left| \frac{Z_1 - Z_2}{Z_1 + Z_2} \right|$$

Reflection coefficient, ρ, given fwd/rev power

$$\rho = \sqrt{\frac{P_{\text{rev}}}{P_{\text{fwd}}}}$$

Return Loss, given forward/reverse power

$$\text{RL(dB)} = 10 \log \left(\frac{P_{\text{fwd}}}{P_{\text{rev}}} \right)$$

Return Loss, given VSWR

$$\text{RL(dB)} = -20 \log \left(\frac{\text{VSWR} - 1}{\text{VSWR} + 1} \right)$$

Return Loss, given reflection coefficient

$$\text{RL(dB)} = -20 \log (\rho)$$

Mismatch Loss given forward/reverse power

$$\text{ML(dB)} = 10 \log \left(\frac{P_{\text{fwd}}}{P_{\text{fwd}} - P_{\text{rev}}} \right)$$

Mismatch Loss, given reflection coefficient

$$ML(dB) = -10 \log \left(1 - \rho^2\right)$$

E-Field from Differential-Mode Current

$$|E_{D,\max}| = 2.63 \times 10^{-14} \frac{|I_D| f^2 L S}{d}$$

I_D = differential-mode current in loop (A)
F = frequency (Hz)
L = length of loop (m)
S = spacing of loop (m)
d = measurement distance (3m or 10m, typ.)

(Assumption that the loop is electrically small and measured over a reflecting surface)

E-Field from Common-Mode Current

$$|E_{C,\max}| = 1.257 \times 10^{-6} \frac{|I_C| f L}{d}$$

I_C = common-mode current in wire (A)
F = frequency (Hz)
L = length of wire (m)
d = measurement distance (3m or 10m, typ.)

(Assumption that the wire is electrically short)

Resonance (rectangular enclosure)

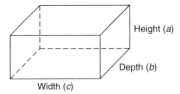

FIGURE 22 A rectangular enclosure can resonate if excited.

$$(f)_{mnp} = \frac{1}{2\sqrt{\epsilon\mu}} \sqrt{\left(\frac{m}{a}\right)^2 + \left(\frac{n}{b}\right)^2 + \left(\frac{p}{c}\right)^2}$$

Where: ϵ = material permittivity, μ = material permeability and m, n, p are integers and a, b, c are the height, depth and width.

Cavity resonance can only exist if the largest cavity dimension is greater, or equal, to one-half the wavelength. Below this cutoff frequency, cavity resonance cannot exist. In this configuration (where a < b < c), the TE_{011} mode is dominant, because it occurs at the lowest frequency at which cavity resonance can exist.

Resonance (circular enclosure)

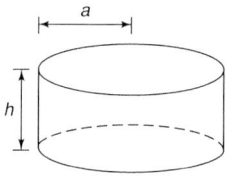

FIGURE 23 Circular resonant cavity.

For

$$a \geq h/2.03$$

For a circular cavity

$$f = 2.4049/2\pi a \sqrt{\mu_0 \varepsilon_0}$$

Example: for $a = 9$ cm and $h = 6$ cm (a typical cookie tin):

$$f = \frac{2.4049}{2\pi (0.09) \sqrt{(8.854 \times 10^{-12})(4\pi \times 10^{-7})}}$$

$$f = 1,275 \text{ MHz}$$

Antenna (Far Field) Relationships

Gain, dBi to numeric

$$\text{Gain}_{\text{numeric}} = 10^{(\text{dBi}/10)}$$

Gain, numeric to dBi

$$\text{dBi} = 10 \log (\text{Gain}_{\text{numeric}})$$

Antenna Factor Equations (based on a 50 Ω system and using polarization matched antennas)

$$\text{AF(dB/m)} = \text{E(dB}\mu\text{V/m)} - \text{Vr(dB}\mu\text{V)}$$

Where:

AF = Is the antenna factor of the measuring antenna (dB/m)
E = Field strength at the antenna in dBμV/m
Vr = Output voltage from receiving antenna in dBμV

Gain (dBi) to Antenna Factor

$$\text{AF} = 20 \log[\text{f(MHz)}] - \text{G(dBi)} - 29.79 \text{ dB}$$

Field Strength given Antenna Factor and spectrum analyzer output adjusted for system losses (V_o)

$$\text{E(dB}\mu\text{V/m)} = \text{AF(dB/m)} + \text{V}_o\text{(dB}\mu\text{V)}$$

Where:

AF is the antenna factor of the measuring antenna (dB/m)
E is the unknown or measured electric field strength

$\mathbf{V}_\text{o}$ is the adjusted spectrum analyzer output (calibrated for cable & connector, or system, losses)

Field strength given Power (watts), antenna gain (numeric) and distance (meters)

$$V/m = \frac{\sqrt{30PG}}{d}$$

Field strength, given Power (watts), antenna gain (dBi) and distance (meters)

$$V/m = \frac{\sqrt{30P\,10^{(\text{dBi}/10)}}}{d}$$

Transmit power required, given desired V/m, antenna gain (numeric) and distance (meters)

$$P_\text{transmit} = \frac{((V/m)d)^2}{30G}$$

Transmit power required, given desired V/m, antenna gain (dBi) and distance (meters)

$$P_\text{transmit} = \frac{((V/m)d)^2}{30(10^{(\text{dBi}/10)})}$$

Wave Impedance versus Distance

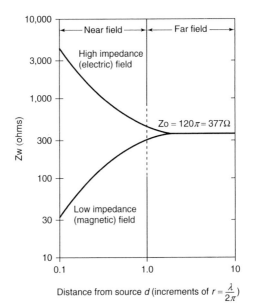

FIGURE 24 Variation of wave impedance versus distance from the source.

The near-field far-field transition occurs at about 1/6 wavelength.

E-Field Levels versus Transmitter Pout

Pout (W)	V/m at 1 m	V/m at 3 m	V/m at 10 m
1	5.5	1.8	0.6
5	12.3	4.1	1.2
10	17.4	5.8	1.7
25	27.5	9.2	2.8
50	38.9	13.0	3.9
100	55.0	18.3	5.5
1000	173.9	58.0	17.4

Assuming the antenna gain is 1.

Device	Approx Freq	Max Power	Approx V/m at 1 m
Citizens Band	27 MHz	5 W	12
FRS	465 MHz	500 mW	4
GMRS	465 MHz	1 to 5 W	5.5 to 12
3G Mobile Phone	830 MHz / 1.8 GHz	400 mW	3.5

These license-free devices may be used to determine radiated immunity of an EUT by holding them close to the EUT or I/O or power cable. GMRS radios require a license.

■ Miscellaneous Information

Galvanic Series

The galvanic and triboelectric series are often confused but are useful for different purposes.

The galvanic series allows us to determine which of two metals in an electrolyte (or some other reactive environment) will experience galvanic corrosion, with the less noble (sacrificial) metal protecting the other, more noble metal. This is also known as cathodic protection. This protection is dependent upon the electrolyte, surface finish and material composition. It is important to understand bonding and corrosion of enclosures, especially in harsh environments.

Cathode	Anode										
	Magnesium	Zinc	Aluminum	Cadmium	Tin	Iron, steel	Chromium	Brass	Copper, bronze	Nickel, monel	Stainless steel
Zinc	.075										
Aluminum	1.05	.029									
Cadmium	1.05	.029	0.01								
Tin	1.36	.060	0.31	0.31							
Iron, steel	1.30	.029	0.32	0.32	0.01						
Chromium	1.39	0.65	0.34	0.34	0.03	0.02					
Brass	1.54	0.78	0.50	0.50	0.22	0.20	0.02				
Copper/bronze	1.58	0.82	0.55	0.55	0.24	0.23	0.11	0.02			
Nickel/monel	1.58	0.82	0.56	0.56	0.25	0.25	0.12	0.03	0.01		
Stainless steel	1.67	0.91	0.64	0.64	0.35	0.32	0.20	0.11	0.02	0.08	
Silver	1.78	1.02	0.75	0.75	0.44	0.43	0.31	0.22	0.21	0.19	0.11

1. For units which will be subjected to salt spray or salt water, metal should be chosen where the potential difference is less than 0.25V.

2. Where it is possible the unit will be subjected to high humidity that is not salt laden, then the potential difference should not exceed 0.45V.

FIGURE 25 Galvanic series – minimize the differential voltage.

Triboelectric Series

The triboelectric series determines how different materials will accept or give up electrons when rubbed together, to determine which material becomes more positive and which becomes more negative.

(+) POSITIVE END OF SERIES (lower work function)

Air
Human Skin
Asbestos
Glass
Nylon
Wool
Lead
Silk
Aluminum
Paper
Cotton
Steel
Hard rubber
Epoxy-Fiberglass
Nickel & copper
Brass & silver
Synthetic rubber
Orlon
Saran
Polyethylene
Teflon
Silicone rubber

(−) NEGATIVE END OF SERIES (higher work function)

Actual charge transfer depends on several factors, including specific materials, surface finish, temperature, humidity, etc.

■ Useful Software

PC

LTspice (free from Linear Technology)

Various PC board layout viewers (refer to layout manufacturers)

Macintosh

EasyDraw (for producing schematics, drawings and graphics)

McCAD (reviewing PC board layouts)

MacSpice (spice simulator)

iPad/iPhone

dB Calculator (Rohde & Schwarz)

Field Strength & Power Estimator (Rohde & Schwarz)

Interference Hunter (Rohde & Schwarz)

Aviation RF Link (Rohde & Schwarz)

Wireless Communication Calculator (Rohde & Schwarz)

iCircuit (schematic capture and analysis)

LineCalc (coax cable loss and electrical length)

Electronic TB (multitude of handy electronic design aids)

μWave Calc (Agilent Technologies μW/RF calculator)

PCB Calc (Agilent Technologies microstrip/stripline calculator)

RF Tools (Huber+Suhner RF tools for reflection, frequency/wavelength, signal delay, impedance and dB)

RF Toolbox Pro (a comprehensive collection of RF-related tools and references)

E Formulas (multitude of electronics-related formulas and calculators)

Circuit Lab (analyzes DC/AC linear and non-linear and transient circuits)

dB Calc (calculates/converts dB)

EE Toolkit (another well-done component and circuits reference)

Buyer's Guide (ITEM buyer's guide to EMC products and services)

Directives (listing of EU directives with links to the document)

Calculator Pro (a good EE calculator)

Power One SE Calculator (engineering calculator with equation solver)

Spicy (well-done schematic capture and spice analysis)

Spicy SWAN (a comprehensive Spice and "simulation by wave analysis" is a step up from Spicy in that it focuses more on signal propagation effects. Useful for high-frequency transmission line analysis.

Android

dB Calculator (Rohde & Schwarz)

Field Strength & Power Estimator (Rohde & Schwarz)

Interference Hunter (Rohde & Schwarz)

Aviation RF Link (Rohde & Schwarz)

RF & Microwave Toolbox (by Elektor, a collection of RF tools)

RF Engineering Tools (by Freescale, a compilation of calculators and converters for RF and microwave design)

RF Calc (by NXP, series/parallel, unit conversions, thermal resistance)

RF Calculator (by Lighthorse Tech, wavelength, propagation velocity, etc.)

EMC & Radio Conversion Utility (by TRAC Global, converters, path loss, EIRP, wavelength, etc.)

■ References

Books

ARRL, The RFI Handbook, 3rd edition, 2010.

Eric Bogatin, *Signal Integrity - Simplified*, Prentice Hall PTR, 2004.

Stephen H. Hall, Garrett W. Hall & James A. McCall, *High-Speed Digital System Design: A Handbook of Interconnect Theory and Design Practices*, John Wiley & Sons, 2000.

Howard W. Johnson & Martin Graham, *High-Speed Digital Design: A Handbook of Black Magic*, Prentice Hall PTR, 1993.

Howard W. Johnson & Martin Graham, *High-Speed Signal Propagation: Advanced Black Magic*, Prentice Hall PTR, 2003.

Elya B. Joffe and Kai-Sang Lock, *Grounds for Grounding: A Circuit to System Handbook*, Wiley, 2010.

Kenneth L. Kaiser, *Handbook of Electromagnetic Compatibility*, CRC Press, 2005.

Ralph Morrison, *Grounding and Shielding Circuits and Interference, 5th ed.*, Wiley-Interscience, 2007.

Henry W. Ott, *Electromagnetic Compatibility Engineering*, John Wiley & Sons, 2009.

Clayton R. Paul, *Introduction to Electromagnetic Compatibility, 2nd ed.*, Wiley-Interscience, 2006.

Lee W. Ritchey and John Zasio, *Right the First Time – A Practical Handbook on High Speed PCB and System Design*, Speeding Edge, Volume 1 (2003) and Volume 2 (2006). Available at www.speedingedge.com.

Doug Smith, *High Frequency Measurements and Noise in Electronic Circuits*, Van Nostrand 1993.

Steven H. Voldman, *ESD: Physics and Devices*, John Wiley & Sons, Ltd, 2004.

Würth Electronics, Brander, T. et al., *Trilogy of Magnetics: Design Guide for EMI Filter Design, SMPS and RF Circuits, 4th ed.*, Würth Elektronik eiSos GmbH & Co., 2010.

Brian Young, *Digital Signal Integrity: Modeling and Simulation with Interconnects and Packages*, Prentice Hall PTR, 2001.

EMC Magazines

InCompliance Magazine (www.incompliancemag .com)

Interference Technology (www.interferencetechnology.com)

IEEE Electromagnetic Compatibility Magazine, published by the IEEE EMC Society, (www.emcs.org/newsletters.html)

Electromagnetic News Report (www.7ms.com)

Safety & EMC, China (www.semc.cesi.cn)

Compliance Engineering (stopped publication in 2006, but still available at www.ce-mag.com)

The EMC Journal, UK, (www.compliance-club .com)

EMC Organizations

Automotive EMC (www.autoemc.net)

IEEE EMC Society (www.ewh.ieee.org/soc/emcs/)

ESD Association (www.esd.org)

iNARTE (International Association for Radio, Telecommunications and Electromagnetics) (www.narte.org)

EMC Standards Organizations

ANSI (American National Standards Institute) (www.ansi.org)

ANSI Accredited C63 (www.c63.org)

IEEE Standards Association (www.standards .ieee.org)

IEEE EMC Society Standards Development Committee (SDCOM) (http://www.emcs.org/standards/index.html)

SAE (Society of Automotive Engineers) (www.sae.org/)

SAE EMC Standards Committee (www.sae.org/standards/)

EMCIA (Electromagnetic Compatibility Industry Association, UK) (www.emcia.org)

IEC (International Electrotechnical Commission) (www.iec.ch/index.htm)

CISPR (International Special Committee on Radio Interference) (http://www.iec.ch/emc/iec-emc/iec-emc-players-cispr.htm)

APLAC (Asia Pacific Laboratory Accreditation Cooperation) (www.aplac.org)

CSA (Canadian Standards Association) (www.csa.ca)

FCC (Federal Communications Commission, US) (www.fcc.gov)

IBIS (Input/Output Buffer Specification) (www.eigroup.org/ibis/default.html)

ISO (International Organization for Standards) (www.iso.org/iso/home.html)

VCCI (Voluntary Control Council for Interference, Japan) (www.vcci.jp/vcci e/)

LinkedIn Groups

LinkedIn is the world's largest professional's networking and discussion forum. Several discussion groups related to EMC are listed here. Create your free profile at: www.linkedin.com.

- Aircraft and Spacecraft ESD/EMI.EMC Issues
- Automotive EMC Troubleshooting Experts
- EMC – Electromagnetic Compatibility
- EMC Experts
- EMC Jobs
- EMC Testing and Compliance
- EMI and EMC Consultants
- Electromagnetic Compatibility Automotive Group
- Electromagnetic Compatibility Forum
- Electromagnetics and Spectrum Engineering Group
- ESD Experts
- iNARTE
- Military EMC Forum
- RTCA/DO-160 Experts

Common Symbols

Å	Angstrom, unit of length, one ten billionth of a meter
A	Amperes, unit of electrical current
AC	Alternating Current

AM	Amplitude modulated
cm	Centimeter, one hundredth of a meter
dBm	dB with reference to 1 mW
dBµA	dB with reference to 1 µA
dBµV	dB with reference to 1 µV
DC	Direct Current
E	"E" is the electric field component of an electromagnetic field.
E/M	Ratio of the electric field (E) to the magnetic field (H), in the far-field this is the characteristic impedance of free space, approximately 377 Ω
EM	Electromagnetic
EMC	Electromagnetic compatibility
EMI	Electromagnetic Interference
FM	Frequency modulated
GHz	Gigahertz, one billion Hertz (1,000,000,000 Hertz)
H	"H" is the magnetic field component of an electromagnetic field.
Hz	Hertz, unit of measurement for frequency
I	Electric current
kHz	Kilohertz, one thousand Hertz (1000 Hertz)
λ	Lambda, symbol for wavelength, distance a wave travels during the time period necessary for one complete oscillation cycle

MHz	Megahertz, one million Hertz (1,000,000 Hertz)
μm	Micrometer, unit of length, one millionth of an meter (0.000001 meter)
m	Meter, the fundamental unit of length in the metric system
mil	Unit of length, one thousandth of an inch
mW	Milliwatt (0.001 Watt)
mW/cm^2	Milliwatts per square centimeter (0.001 Watt per square centimeter area), a unit for power density, one mW/cm^2 equals ten W/m^2
P_d	Power density, unit of measurement of power per unit area (W/m^2 or mW/cm^2)
R	Resistance
RF	Radio Frequency
RFI	Radio Frequency Interference
V	Volts, unit of electric voltage potential
V/m	Volts per meter, unit of electric field strength
W/m^2	Watts per square meter, a unit for power density, one W/m^2 equals 0.1 mw/cm^2
Ω	Ohms, unit of resistance

Ref: ANSI/IEEE 100-1984, IEEE Standard Dictionary of Electrical and Electronics Terms, 1984.

EMC Acronyms

AF (Antenna Factor) – The ratio of the received field strength to the voltage at the terminals of a receiving antenna. Units are 1/m.

ALC (Absorber-Lined Chamber) – A shielded room with RF-absorbing material on the walls and ceiling. In many cases, the floor is reflective.

AM (Amplitude Modulation) – A technique for putting information on a sinusoidal carrier signal by varying the amplitude of the carrier.

BCI (Bulk Current Injection) – An EMC test where common-mode currents are coupled onto the power and communications cables of an EUT.

CE (Conducted Emissions) – The RF energy generated by electronic equipment, which is conducted on power cables.

CE Marking – The marking signifying a product meets the required European Directives.

CENELEC – French acronym for the "European Committee for Electrotechnical Standardization".

CI (Conducted Immunity) – A measure of the immunity to RF energy coupled onto cables and wires of an electronic product.

CISPR – French acronym for "Special International Committee on Radio Interference".

Conducted – Energy transmitted via cables or PC board connections.

Coupling Path – a structure or medium that transmits energy from a noise source to a victim circuit or system.

CS (Conducted Susceptibility) – RF energy or electrical noise coupled onto I/O cables and power wiring that can disrupt electronic equipment.

CW (Continuous Wave) – A sinusoidal waveform with a constant amplitude and frequency.

EMC (Electromagnetic Compatibility) – The ability of a product to coexist in its intended electromagnetic environment without causing or suffering disruption or damage.

EMI (Electromagnetic Interference) – When electromagnetic energy is transmitted from an electronic device to a victim circuit or system via radiated or conducted paths (or both) and which causes circuit upset in the victim.

EMP (Electromagnetic Pulse) – Strong electromagnetic transients such as those created by lightning or nuclear blasts.

ESD (Electrostatic Discharge) – A sudden surge in current (positive or negative) due to an electric spark or secondary discharge causing circuit disruption or component damage. Typically characterized by rise times less than 1 ns and total pulse widths on the order of microseconds.

EU – European Union.

EUT (Equipment Under Test) – The device being evaluated.

Far Field – When you get far enough from a radiating source the radiated field can be considered planar (or plane waves).

FCC – U.S. Federal Communications Commission

FM (Frequency Modulation) – A technique for putting information on a sinusoidal "carrier" signal by varying the frequency of the carrier.

IEC – International Electrotechnical Commission

ISM (Industrial, Scientific and Medical equipment) – A class of electronic equipment including industrial controllers, test & measurement equipment, medical products and other scientific equipment.

ITE (Information Technology Equipment) – A class of electronic devices covering a broad range of equipment including computers, printers and external peripherals; also includes telecom-munications equipment, and multi-media devices.

LISN (Line Impedance Stabilization Network) – Used to match the 50-Ohm impedance of measuring receivers to the power line.

Near Field – When you are close enough to a radiating source that its field is considered spherical rather than planar.

Noise Source – A source that generates an electromagnetic perturbation or disruption to other circuits or systems.

OATS (Open Area Test Site) – An outdoor EMC test site free of reflecting objects except a ground plane.

PDN (Power Distribution Network) – The wiring and circuit traces from the power source to the electronic circuitry. This includes the parasitic components (R, L, C) of the circuit board, traces, bypass capacitance and any series inductances.

PLT (Power Line Transient) – A sudden positive or negative surge in the voltage on a power supply input (DC source or AC line).

Radiated – Energy transmitted through the air via antenna or loops.

RFI (Radio Frequency Interference) – The disruption of an electronic device or system due to electromagnetic emissions at radio frequencies (usually a few kHz to a few GHz). Also EMI.

RE (Radiated Emissions) – The energy generated by a circuit or equipment, which is radiated directly from the circuits, chassis and/or cables of equipment.

RI (Radiated Immunity) – The ability of circuits or systems to be immune from radiated energy coupled to the chassis, circuit boards and/or cables. Also Radiated Susceptibility (RS).

RF (Radio Frequency) – A frequency at which electromagnetic radiation of energy is useful for communications.

RS (Radiated Susceptibility) – The ability of equipment or circuits to withstand or reject nearby radiated RF sources. Also Radiated Immunity (RI).

SSCG (Spread Spectrum Clock Generation) – This technique takes the energy from a CW clock signal and spreads it out wider, which results in a lower effective amplitude for the fundamental and high-order harmonics. Used to achieve improved radiated or conducted emission margin to the limits.

SSN (Simultaneous Switching Noise) – Fast pulses that occur on the power bus due to switching transient currents drawn by the digital circuitry.

TEM (Transverse Electromagnetic) – An electromagnetic plane wave where the electric and magnetic fields are perpendicular to each other everywhere and both fields are perpendicular to the direction of propagation. TEM cells are often used to generate TEM waves for radiated immunity (RI) testing.

Victim – An electronic device, component or system that receives an electromagnetic disturbance, which causes circuit upset.

VSWR (Voltage Standing Wave Ratio) – A measure of how well the load is impedance matched to its transmission line. This is calculated by dividing the voltage at the peak of a standing wave by the voltage at the null in the standing wave. A good match is less than 1.2:1.

XTALK (Crosstalk) – A measure of the electromagnetic coupling from one circuit to another. This is a common problem between one circuit trace and another.

NOTES